William Barton Hopkins

On the organization and absorption of sterilized dead bone dowels

William Barton Hopkins

On the organization and absorption of sterilized dead bone dowels

ISBN/EAN: 9783337873783

Printed in Europe, USA, Canada, Australia, Japan

Cover: Foto ©berggeist007 / pixelio.de

More available books at **www.hansebooks.com**

On the Organization and Absorption of Sterilized Dead Bone Dowels

BY

WILLIAM BARTON HOPKINS, M.D.,

SURGEON TO THE EPISCOPAL HOSPITAL AND SURGEON TO OUT-
DEPARTMENT OF PENNSYLVANIA HOSPITAL.

AND BY

CHARLES B. PENROSE, M.D., PH D.,

ASSISTANT SURGEON TO THE HOSPITAL OF THE UNIVERSITY OF
PENNSYLVANIA, SURGEON TO OUT-DEPARTMENT
OF PENNSYLVANIA HOSPITAL

*Read in the Section of Surgery and Anatomy, at the Fortieth An-
nual Meeting of the American Medical Association, June, 1889.*

Reprinted from the "Journal of the American Medical
Association," April 5, 1890.

CHICAGO
PRINTED AT THE OFFICE OF THE ASSOCIATION
1890.

ON THE ORGANIZATION AND ABSORPTION OF STERILIZED DEAD BONE DOWELS.

Mr. Wm. Scovell Savory, in 1864, demonstrated, by a series of experiments performed upon various animals, principally the rabbit, that ivory pegs driven into healthy living bone would, after a time, undergo greater or less absorption. He found, however, that in order to obtain this result, it was necessary that the peg be driven tightly into a hole which was a trifle too small for it. Upon this observation, he concluded that firm pressure upon the surface of dead bone against the living was the essential factor in causing the absorption. His experiments were, of course, done long before the antiseptic era, and we may venture to assume that that gentleman to-day shares our opinion, that the dead bone was absorbed when it pressed firmly against the living because the closeness of its contact diminished the space for the accumulation of septic matter. Bone pegs have since been used to fix together extremities of bones after excisions ; but we believe that the experiments here recorded are the first of the kind which have been performed, with antiseptic precautions, upon the lower animals, and with the view of determining how long foreign bone can be depended upon for giving fixation, and the ultimate changes which take place in it during the processes of absorption or organization, when placed in contact with living osseous tissue.

Experiment No. 1.—A large cur puppy was etherized, and after exposure of the shaft of the right femur, a sterilized ox-bone dowel was placed in a hole, which had been drilled transversely through the shaft, and was cut óff flush with the surface. The incision for exposing the bone in

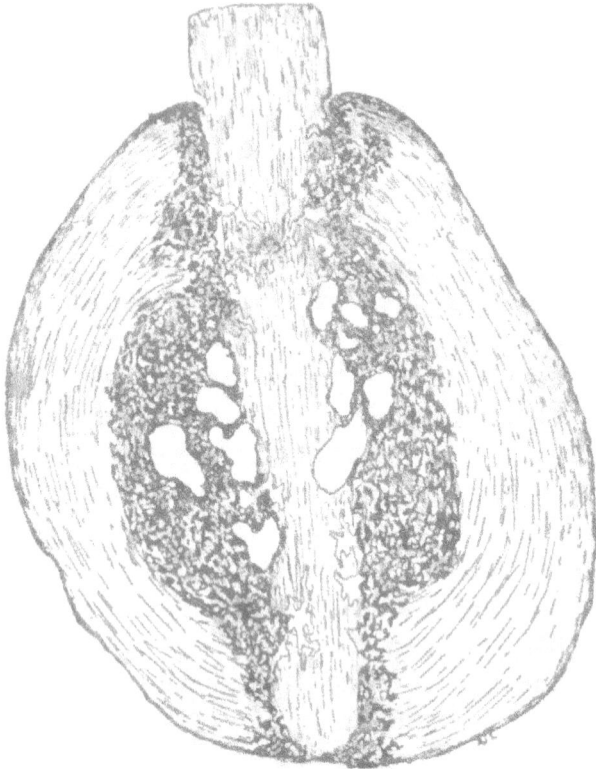

FIGURE I.—Showing bone dowel after five weeks undergoing organization and absorption.

this case, as in all the subsequent experiments, being carried down in the intermuscular septum overlying the linea aspera, little muscular tissue was divided, and hence but little bleeding occurred during the operation. The wound was approximated by a continuous suture of chromicized catgut, and was sealed with collodion and powdered iodoform. The wound healed without suppuration, and at the end of five weeks the dog was again etherized and the section of the femur in which the bone dowel had been placed, was removed. From this, as from other experiments, Dr. Allen J. Smith made careful microscopic sections, the best of which were unfortunately destroyed in the fire at the histological laboratory of the University of Pennsylvania. The remarkable changes which occurred in five weeks are shown in Figure 1, and in the microscopic section, Figure 2. Two distinct phenomena are ap-

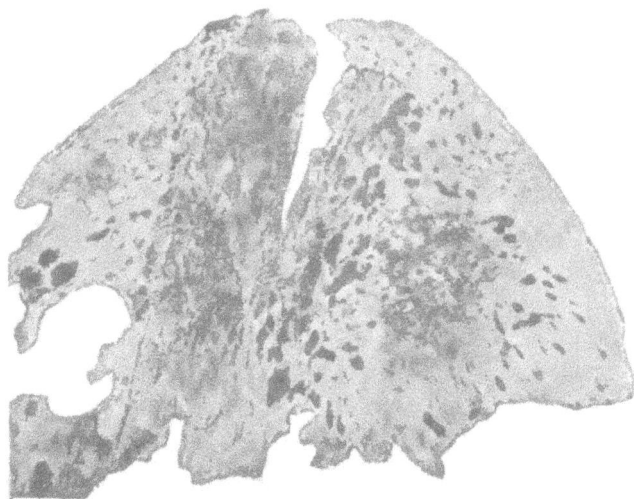

FIGURE 2.—Showing microscopic appearances of Experiment No. 1.

parent : that of organization, and that of absorption. The extremities of the dowel which were in contact with the shaft of the femur had become thoroughly organized, being full of Haversian canals continuous with those of the dog bone, of which they had become part, and on fresh section showing the pink hue of living vascular bone. The intervening portion, between these extremities, corresponding to the medullary cavity, showed no attempt at organization, as it was pure white on fresh section, and contained no Haversian canals, but the erosions on its surface clearly indicated that rapid absorption was taking place.

A deep furrow which surrounds the dowel at the point where it penetrated the periosteum, shows an effort of this membrane to sever that portion of the dowel which projected beyond the level of the shaft of the femur, and would in all probability, had the action gone on, have allowed this free portion to drop off into the surrounding tissues.

Experiment No. 2.—The left femur of a healthy carriage dog was exposed and a dowel an eighth of an inch in diameter was introduced, as in the preceding experiment. At the end of seven weeks the femur at the site of the operation was again sought, and the only evidences that remained of

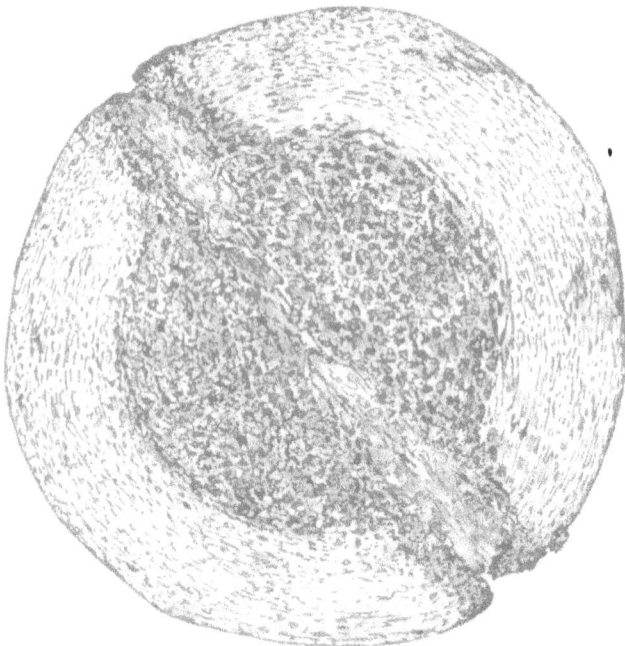

FIGURE 3.—Memory sketch of the appearances of bone dowel after seven weeks.

the introduction of the peg were pin-point dimples in the opposite sides of the shaft of the bone. This section was sent to Dr. Smith unsawn, and we have his statement that organization and absorption had been so complete that almost all vestiges of the dowel had disappeared. All of the microscopic sections from this specimen were destroyed in the fire. Figure 3 represents a memory sketch.

Experiments Nos. 3, 4, 5, 6, 7, 8, 9 and 11 were all done with the view of determining the effect produced by splicing together the fragments of the femur with bone after it had been divided. While the success attained in these experiments was only measurable, they may be considered satisfactory when the inordinate liberties which were taken with the bone are remembered. The simplest form of splicing was done by two transfixing dowels, as would be applicable to the fixation of a simple oblique fracture. A transverse section of the shaft was spliced by the introduction of a dowel, which accurately fitted into the medullary cavity one inch or less on either side of the fracture ; the pin being, in some of the experiments, held in position above and below by cross-pins of a smaller size, made to pass through it and the shaft of the bone. Again, an hiatus in a resected femur was filled up by a medullary splice, retaining the fragments at a distance of half an inch apart. The material of which this medullary pin was composed was either dead bone or freshly removed living dog-bone.

All of these cases suppurated, and at the end of periods varying from nine weeks to nine months the dog was killed and the femur removed. In all, more or less necrosis was found to have occurred, accompanied by different degrees of osteitis, and without organization of the dowels. The best result was obtained in Experiment No. 7.

Three-fourths of an inch (containing in it the
site of a previous experiment) of the femur was
resected and a dowel a quarter of an inch in diameter introduced into the medullary cavity,
which latter had been reamed out with a quarter
inch drill. This joint was made very firm by
eighth-of-an-inch cross pegs, above and below.
In this case a sinus continued for several months,
but finally closed, leaving the dog with a very
strong, useful limb. At the end of nine months
the dog was killed and the femur removed. Firm

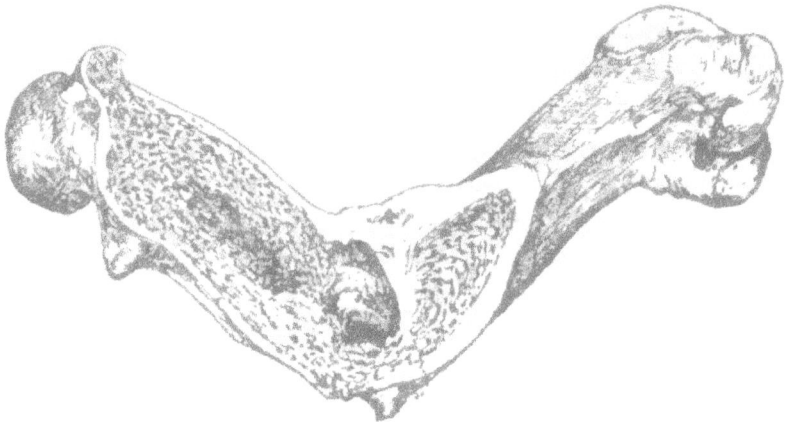

FIGURE 4.—Femur of dog in which dowel filled medullary cavity, after nine months; entire absorption, union with angular deformity.

union, with angular deformity (Figure 4), was
found to have taken place; a few small fragments
of necrosed dog bone, undergoing absorption,
were found on section in a small cavity; but no
signs remained of the heavy medullary dowel, nor
of the small cross pegs.

Experiment No. 10.—Experiment No. 10 was
simply a repetition of experiments Nos. 1 and 2,
the dowel being allowed to remain eight weeks.
The processes of absorption and organization were
found to have gone on, as in the other cases, to
a somewhat more advanced degree. Throughout
its entire length, extension of the blood vessels

of the bone had penetrated the dowel (Figure 5), particularly so at points where it was in continuity with endostium.

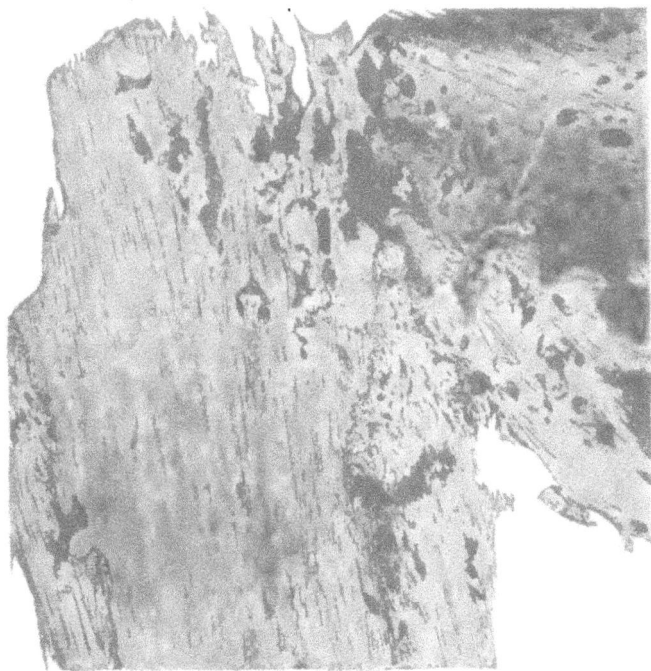

FIGURE 5.—Showing microscopic appearance of bone dowel after eight weeks (the dowel occupying the left side of the field).

Mode of Preparation.—The dowels used in these experiments were made from ordinary beef bone, as it is prepared for use in the arts. They were sawn into square rods, and then turned by *dowel-cutters* of appropriate sizes. After thorough boiling they were placed in an alcoholic solution of corrosive sublimate (1–1000), ready for use. They were not found to swell at all in this solution, and therefore accurately fitted the drill-holes made for them, even after prolonged immersion.

The only cases which have come under our observation in which these dowels were used upon the human subject, were in two osteoplastic re-

sections of the foot, after the method of Wladim-
iroff-Mikulicz, performed, the one by Dr. Ferdi-
nand Gross, at the German Hospital in Philadel-
phia, in 1888, and the other by Dr. W. B. Hopkins,
at the Episcopal Hospital, in 1887. In both these
cases dowels were used to fix the tarsus to the
leg. In both they answered a good purpose, and
gave rise to no subsequent trouble.

From these experiments we base the following
deductions :

First. That where sterilized dead bone is placed,
under favorable circumstances, in contact with
living bone, it undergoes organization. When,
on the other hand, it is acted upon by periosteum
it is absorbed, and when placed in the medullary
cavity, in not too large bulk, organization com-
bined with absorption takes place.

Second. That these processes go on, perhaps,
most actively between the fifth and the eighth
week, and are not necessarily associated with any
inflammatory action.

Third. That therefore, where these dowels are
employed to pin together fragments of bone after
fracture, to fix the extremities of bones after re-
sections, or for any other mechanical purpose in
surgery to which they are adapted, they may be
relied upon to do their work for a period of one
month or six weeks, and hence to give ample
time, as a rule, for union to occur. After this,
their presence being no longer required, they
gradually lose their identity in the surrounding
bone, and disappear.

www.ingramcontent.com/pod-product-compliance
Lightning Source LLC
Chambersburg PA
CBHW022034190326
41519CB00010B/1719